Now my comet is returning again. Some people have the false idea that comets act like shooting stars, flashing across the sky in a second or so. Actually, they creep very slowly. You can see my comet for several months in late 1985 and early 1986. But you'll need to know what's going on and where to look.

 Illustrated by Ken Condon. LC# 84-52261. ISBN 0-933346-41-7.

What Is a Comet?

People used to pay more attention to the night sky than they do now. Before electric lights, the night was really *dark*. So anyone who looked up — even from a big city like London where I lived — could see thousands of stars.

The patterns of the stars, such as the Big Dipper, stayed the same year after year. But every once in a while something new would appear among them: a dim, fuzzy ball of light, often with a long tail. These objects were named *comets,* from the Latin words for "hairy star." When one appeared, it usually stayed visible for weeks as it moved slowly among the stars.

In ancient times people believed comets were evil omens. Their presence in the sky caused fear and panic. The ancient Chinese thought comets brought famine, sickness, and war. The Roman emperor Nero ordered a great slaughter of his rivals when a bright comet appeared, to show that the comet foretold their doom rather than his own. Medieval monks believed comets gave off the sulfurous smell of devils.

A few centuries ago, people began taking a close look at these notions. Smarter folks realized that bad things are always hap-

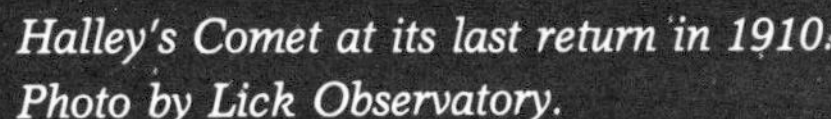

Halley's Comet at its last return in 1910.
Photo by Lick Observatory.

pening somewhere. So when a comet appears, there is always some disaster that superstitious people can blame on the celestial visitor. Astronomers also began watching the motions of comets very carefully and proved that they are far away in space, not just above the clouds as some people thought.

After the telescope was invented around 1610, astronomers got much better looks at comets. They found that all of them have certain things in common. Each has a small, bright core surrounded by a much larger fuzzy head or *coma* (pronounced KO-ma). Streaming away from the head is a fainter *tail.* Very dim comets — those visible only in telescopes — often have no tails at all. But the brightest ones can sprout enormous tails, tens of millions of miles long.

Astronomers noticed that the tails always point away from the Sun. This is true regardless of which way a comet is moving through space. In fact, comets often travel tail first!

Today we know that comets are great clouds of gas and dust, containing pretty much the same kind of stuff we find on Earth. But in a comet this material is very thin, *much* thinner than the air we breathe. Comets, in fact, are mostly empty space. There's practically nothing to them at all!

The only solid part of a comet is its *nucleus,* a cold lump of ice, dust, and stones mixed together with some frozen gases like carbon monoxide and ammonia. The nucleus has been described as a dirty snowball. It is no more than a few miles across — the size of a flying mountain. This is tiny by astronomical standards. But the nucleus creates all the parts of a comet that we see.

A magnified view of Halley's head in 1910.

Here's how it happens. Once in a while,

A comet's solid nucleus giving off gas and dust.

one of these dirty snowballs arrives from deep space, where it's very cold and dark. As the snowball approaches the Sun, it begins to warm up. Its outer parts start to evaporate, and the resulting gas drifts away. Dust that was frozen into the snow is carried along too. As a result, the snowball becomes surrounded by a hazy cloud perhaps a hundred thousand miles across — more than 10 times the size of the Earth. Sunlight makes this cloud shine, and we see it as the comet's head or coma.

The tail is formed by the "wind" of thin gas that always blows outward from the Sun and by the pressure of sunlight itself. These forces push the dust and gas away from a comet's head to form the tail, like wind-blown smoke.

There are probably billions of cold cometary snowballs in the depths of space. But most of them never come our way. Of those that do, most produce such a small cloud that a telescope is needed to see it.

About a dozen such faint comets are discovered every year. Often they are found by amateur astronomers. And since nowadays a comet is named after the person who discovers it, many an amateur using a backyard telescope dreams of finding one that will put his or her name in astronomical history books.

Most comets pass our way only once in thousands or millions of years. They swing by the Sun for their few weeks of glory and then fly back out into the cold and dark, far beyond even the most distant planets. Others return again and again, coming back every few years or every few centuries. Of all such *periodic comets,* Halley is the brightest. And it's certainly the most famous.

Why Is Halley Special?

Let's clear up a couple of things right now. First, please pronounce my name right! "Halley" rhymes with "valley;" *not* with "daily." Got that?

Second — and this may be a disappointment — my comet won't be terribly bright or dramatic. It certainly won't light up the sky. But it does have one big advantage. We know exactly when it's coming and where to look for it.

The main reason Halley's Comet is so special is that it has such a long history. It's been seen for thousands of years and has left its mark on human cultures the world over. The first really clear record of it is from 240 B.C., and it has never been missed on any of its 29 returns since then.

People believed Halley's Comet foretold the death of the Roman ruler Agrippa in 11 B.C., the destruction of Jerusalem in 66 A.D., and the defeat of Attila the Hun in 451. When the comet appeared in 1066, it was regarded as an omen of the Norman

A 1906 cartoon pokes fun at astronomers trying to make sense of a comet.

invasion of England later that year. Its last appearance, in 1910, aroused enormous public interest. That year there were comet parties, comet pills, comet coffee, comet soap — the list goes on and on. Since then, three generations have waited for the return of my comet. Now it's finally happening!

Viewing the Comet

"All right," you say, "that stuff is nice to know, but how do I *see* Halley's Comet? What will it look like?"

That depends on where you are. If you're in the middle of a big city, you won't be able to see it at all. The lights are too bright, and their glare blots out all but the brightest stars and planets — comets too! In smaller towns or suburbs, where the sky is somewhat darker, your chances are better. But to see the comet really well, you'll have to go out to the country where the sky is very dark. A 30-mile drive from most big cities should put enough distance between you and the lights.

Unfortunately, the 1985-86 return of Halley is not one of the best. The comet will not come very close to Earth this time, so it will not appear as big as in the past. And when it's brightest, in April, 1986, it will be best seen from the tropics and the Southern Hemisphere. So in the United States, we'll just have to make the most of the view we've got.

The timetable on the next two pages tells the dates you can see Halley's Comet and what it will look like.

The charts on pages 14 to 19 will show exactly where to find it — even if you don't know a thing about astronomy and have never tried to find anything in the sky before.

For now, here's some general advice.

• First, the sky has to be clear! Even thin clouds will blot out such a dim, wispy thing as the comet. Be especially wary of clouds near the horizon, where it will be when brightest.

• Don't expect much on nights around full moon. Bright moonlight washes out faint things in the sky.

• Find a good, dark location with a wide-open view of the part of the sky you want to see. (We'll get to that shortly.)

• Give your eyes time to adapt to the dark. Arrive plenty early, look at the stars, relax and enjoy the night.

Halley's Comet Timetable

(For North America, southern Europe, and elsewhere at mid-northern latitudes.)

Before August, 1985. Still very far away, Comet Halley is extremely faint. The world's largest telescopes will photograph it as a vague smudge of light.

August-September, 1985. By now the comet is getting bright enough for experienced amateur astronomers with large telescopes and detailed star charts to begin picking it up.

October, 1985. Late this month, when moonlight is no longer a problem, the comet should be widely spotted by amateurs using small telescopes.

November, 1985. Growing steadily brighter, Halley can now be seen in binoculars. On Friday and Saturday nights, Nov. 15 and 16, it passes just south of the Pleiades star cluster. This hazy little cluster, also called the "Seven Sisters," is easy to see with the naked eye in the eastern sky after darkness falls. On

these two evenings, the comet is about half a binocular field of view (as shown in the drawing) to the Pleiades' right or lower right. It's within one binocular field of the Pleiades on the nights of Nov. 13-18. The comet will be a dim, fuzzy blob with no tail. Don't miss this early chance to find it!

December, 1985. Halley becomes barely visible to the naked eye under ideal (very dark) conditions. Binoculars will give a somewhat better view. The comet is high in the south portion of the sky, and if you picked it up near the Pleiades, you should be able to follow it night after night without losing it. Our sky charts (pages 14 through 19) begin December 15th.

January, 1986. The comet brightens slowly, but each night after dusk it is lower and lower in the western sky. By the end of the month it sets before the sky becomes dark, and so it's lost to view.

February, 1986. Halley can't be seen most of this month; it's behind the Sun as seen from the Earth. But during the last week of February it reappears in the morning twilight in the east, with a bright, starlike head and possibly a long tail.

March, 1986. The view gets better at last. By the 20th, before moonlight becomes a problem, Halley should sport an excellent long tail in the southeastern sky just before the first light of dawn. The early bird catches the comet!

April, 1986. Halley is at its best! Toward the end of the first week of April, when moonlight ceases to be a problem, the comet appears as bright as it will get. Unfortunately, it's low in the southern sky before dawn and descending rapidly toward the horizon. You'll need a very good view of the southern horizon, with no trees or buildings in the way. (Find a good hilltop!) This is the time when many people will take tours to the Southern Hemisphere to see Halley in all its glory.

For the next week or so, the comet is out of sight from northern latitudes. Then for the last half of April it returns to the evening sky after dusk, fading and shrinking.

May, 1986, and after. Halley departs into deep space once again. High in the evening sky, it can be followed with binoculars through May and with telescopes until early August. No one will set eyes on it again until around the year 2061.

How To Use the Charts

Most of the time my comet won't be bright enough to catch your eye. So you'll have to know just where to look for it. The charts on the next few pages show exactly where it will be in the sky. Never mind that you don't know a thing about astronomy charts! Just follow these step-by-step instructions.

1. First, hold your hand flat out in front of you at arm's length with your fingers together, like I'm doing here. Close one eye and sight past your hand toward the sky. The width of your hand (four fingers together) is a simple way to measure distance between points in the sky.

Try measuring the size of things in "hands." In the picture, the two stars are three hands apart. Twelve hands should take you a quarter of the way around the horizon. Easy, isn't it?

2. The next step is to find where north is. One way is to use a map. (Don't use a compass; they're not very accurate.) I like to find North by the North Star. As shown in the drawing at lower right, the Pointer stars of the Big Dipper always point to the North Star, which is always due north.

3. Now look at the charts on the next six pages. They're in pairs for different months. The top ones are for the northern United States and southern Canada (or elsewhere near 40° north latitude). The bottom ones are for the southern United States (or elsewhere near 30° north latitude).

Find the right chart for your location and date. Now find the comet's position on that date and mark it with an ×. This is where the comet will be at the time listed on the chart. You should plan to go out and look within about 15 minutes of this time. (If need be, you can find sunrise or sunset times in your local newspaper.)

On the chart, the scales along the bottom and sides indicate hands. Count how many hands the comet is from west, south,

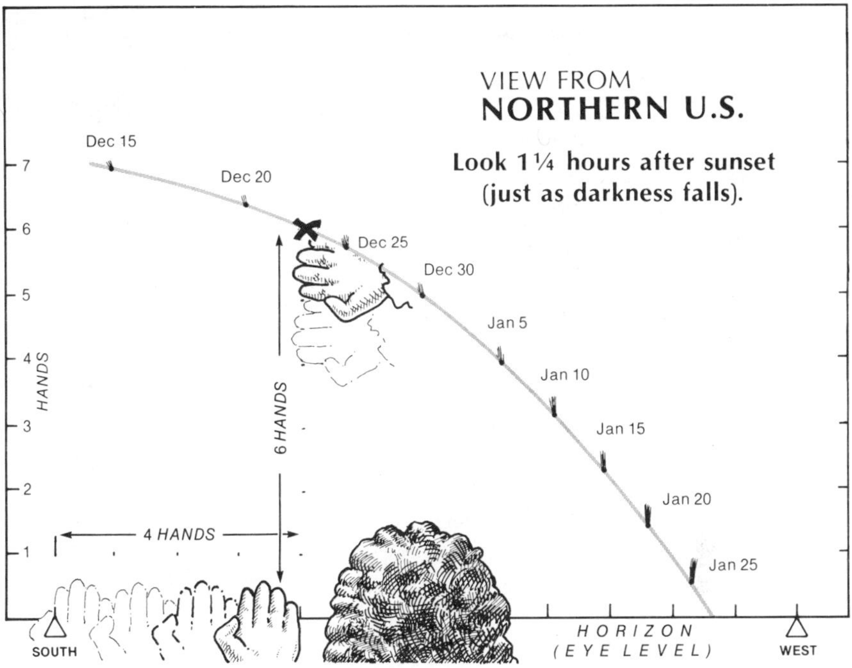

VIEW FROM
NORTHERN U.S.
Look 1¼ hours after sunset
(just as darkness falls).
Dec 15
Dec 20
Dec 25
Dec 30
Jan 5
Jan 10
Jan 15
Jan 20
Jan 25
HANDS
7
6
5
4
3
2
1
6 HANDS
4 HANDS
SOUTH
HORIZON
(EYE LEVEL)
WEST

or east (whichever is nearest) and how many hands it is above the horizon. Be sure to start from the "true horizon," or eye level. This means sighting horizontally and counting up from there, regardless of buildings, trees, or other objects in the way.

Look at the example above. Here the comet is shown for December 23rd. As I am demonstrating, it is 4 hands to the right of south and 6 hands above the horizon.

4. Now find this spot in the sky. The comet will be there. If it's not visible right away, scan the general area carefully with binoculars.

Finding
North
Pointers
APRIL
EVENINGS
North
Star
Pointers
FEBRUARY
EVENINGS
DECEMBER
EVENINGS
Pointers

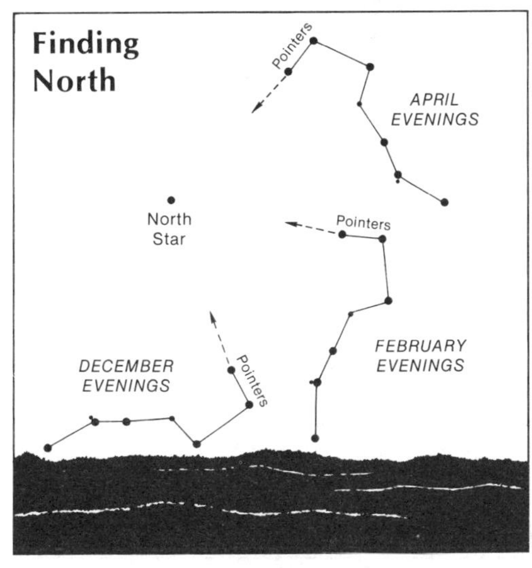

The Big Dipper and the North Star (as seen from the northern U. S.).

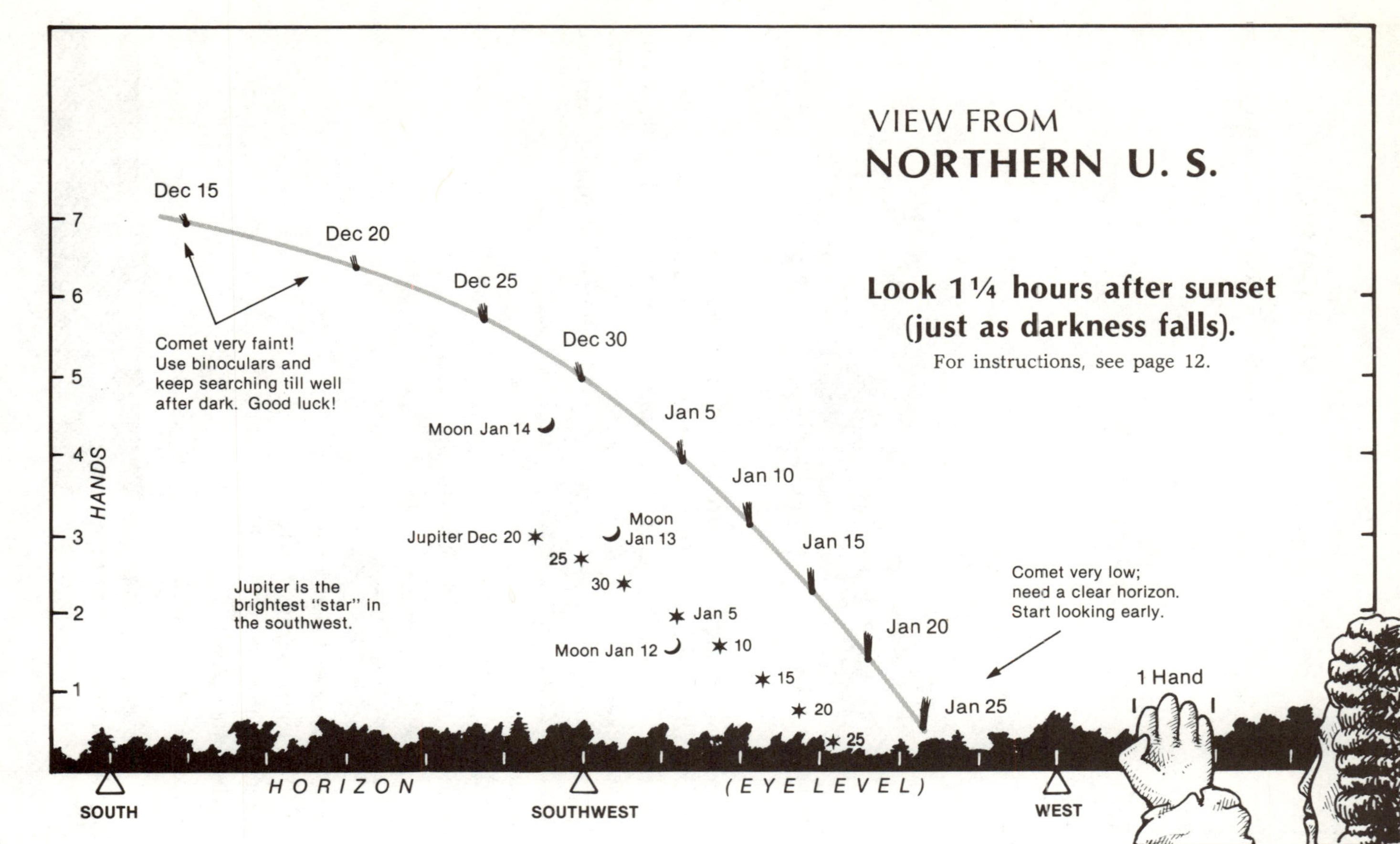
VIEW FROM
NORTHERN U. S.
Look 1¼ hours after sunset
(just as darkness falls).
For instructions, see page 12.
Dec 15
Dec 20
Dec 25
Dec 30
Jan 5
Jan 10
Jan 15
Jan 20
Jan 25
Comet very faint!
Use binoculars and
keep searching till well
after dark. Good luck!
Moon Jan 14
Moon
Jan 13
Jupiter Dec 20
25
30
Jan 5
10
15
20
25
Moon Jan 12
Jupiter is the
brightest "star" in
the southwest.
Comet very low;
need a clear horizon.
Start looking early.
1 Hand
HANDS
7
6
5
4
3
2
1
HORIZON
(EYE LEVEL)
SOUTH
SOUTHWEST
WEST

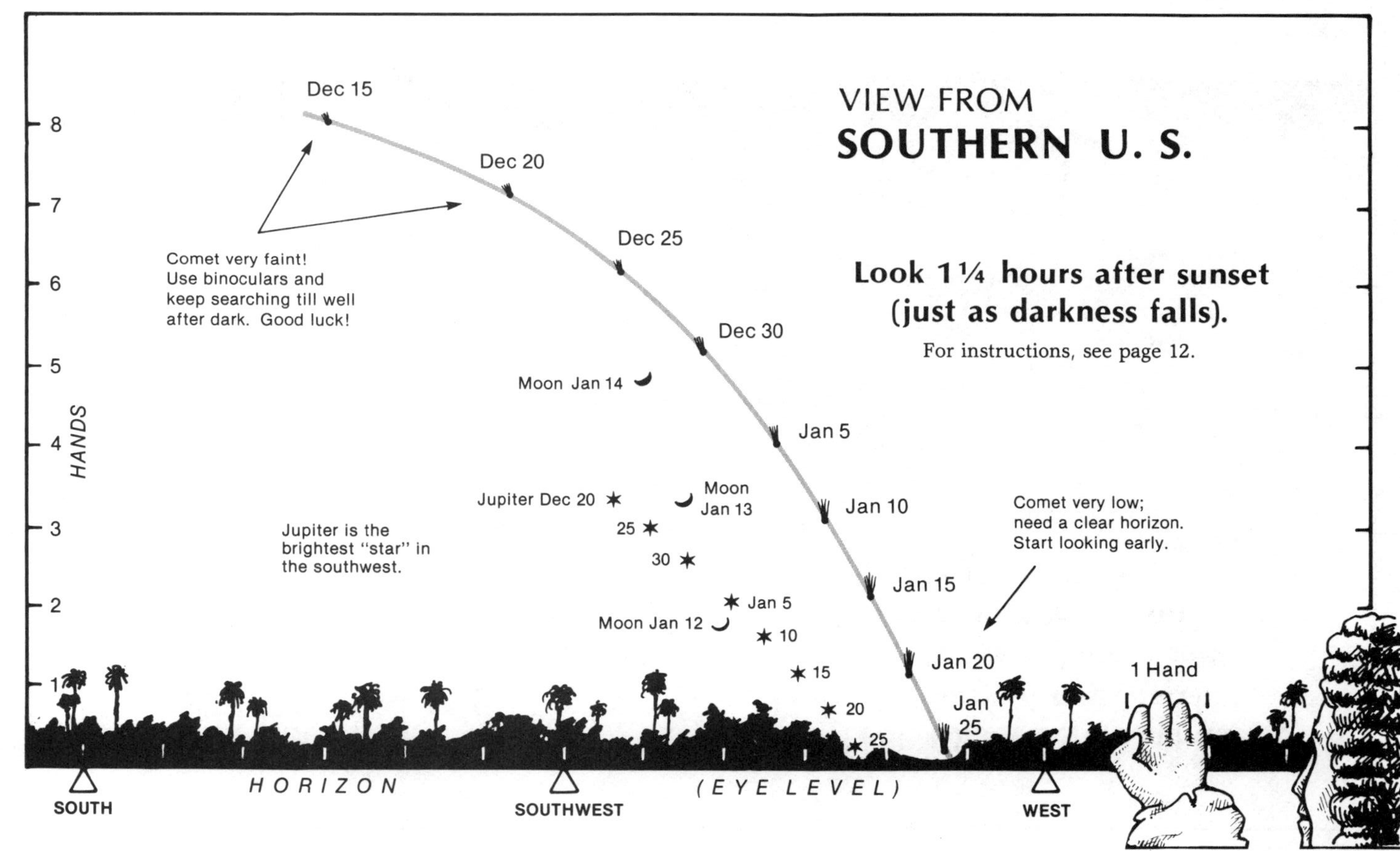
VIEW FROM
SOUTHERN U. S.
Look 1 ¼ hours after sunset
(just as darkness falls).
For instructions, see page 12.
Dec 15
Dec 20
Dec 25
Dec 30
Jan 5
Jan 10
Jan 15
Jan 20
Jan
25
Comet very faint!
Use binoculars and
keep searching till well
after dark. Good luck!
Moon Jan 14
Jupiter Dec 20
25
30
Moon
Jan 13
Jan 5
Moon Jan 12
10
15
20
25
Jupiter is the
brightest "star" in
the southwest.
Comet very low;
need a clear horizon.
Start looking early.
1 Hand
HANDS
8
7
6
5
4
3
2
1
HORIZON
(EYE LEVEL)
SOUTH
SOUTHWEST
WEST

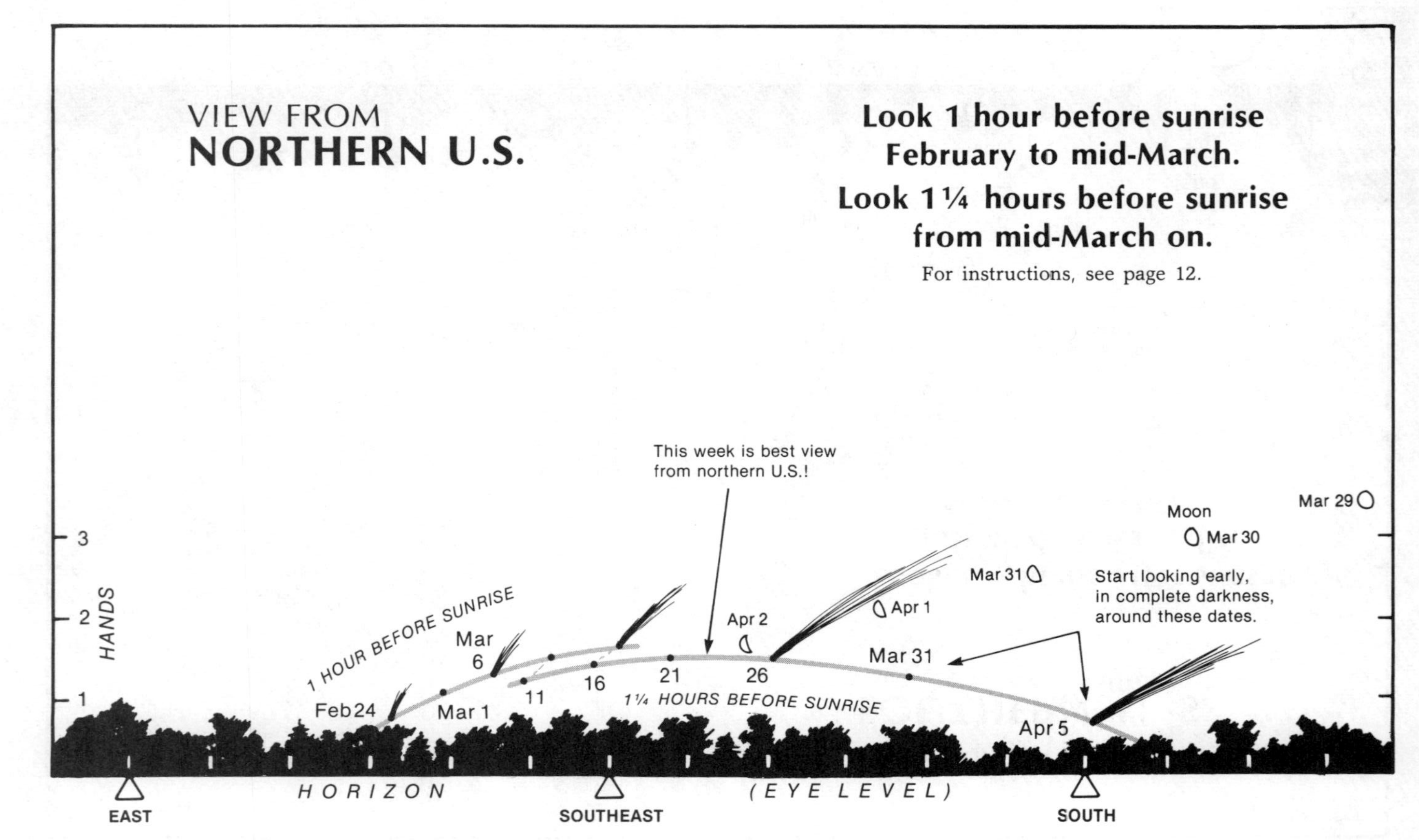
VIEW FROM
NORTHERN U.S.
Look 1 hour before sunrise
February to mid-March.
Look 1¼ hours before sunrise
from mid-March on.
For instructions, see page 12.
This week is best view
from northern U.S.!
Moon
Mar 29
Mar 30
Mar 31
Apr 1
Apr 2
Start looking early,
in complete darkness,
around these dates.
HANDS
3
2
1
1 HOUR BEFORE SUNRISE
Feb 24
Mar 1
Mar
6
11
16
21
26
Mar 31
Apr 5
1¼ HOURS BEFORE SUNRISE
HORIZON
(EYE LEVEL)
EAST
SOUTHEAST
SOUTH

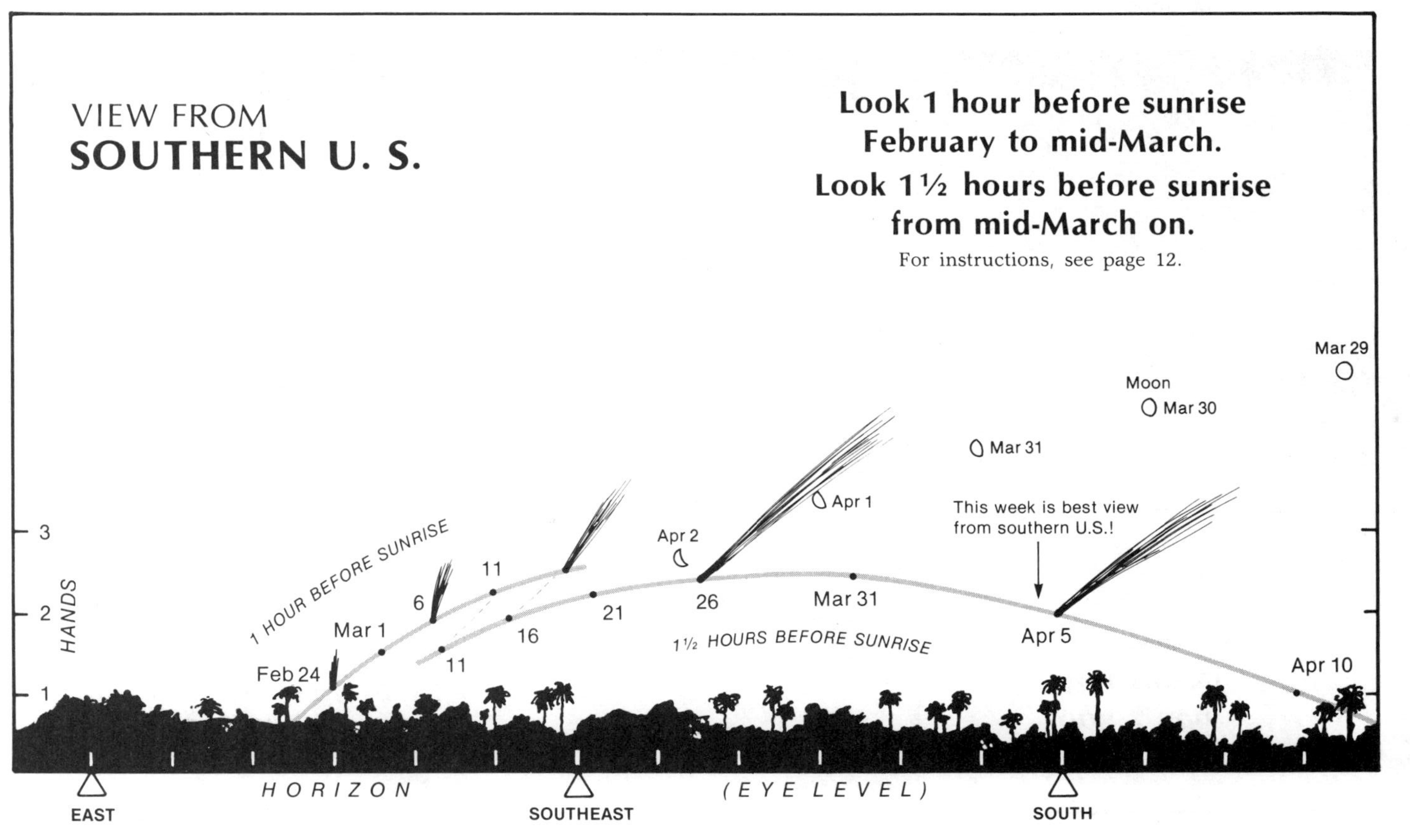
VIEW FROM
SOUTHERN U. S.
Look 1 hour before sunrise
February to mid-March.
Look 1½ hours before sunrise
from mid-March on.
For instructions, see page 12.
Mar 29
Moon
Mar 30
Mar 31
Apr 1
Apr 2
This week is best view
from southern U.S.!
3
2
1
HANDS
1 HOUR BEFORE SUNRISE
11
6
Mar 1
Feb 24
11
16
21
26
Mar 31
1½ HOURS BEFORE SUNRISE
Apr 5
Apr 10
EAST
HORIZON
SOUTHEAST
(EYE LEVEL)
SOUTH

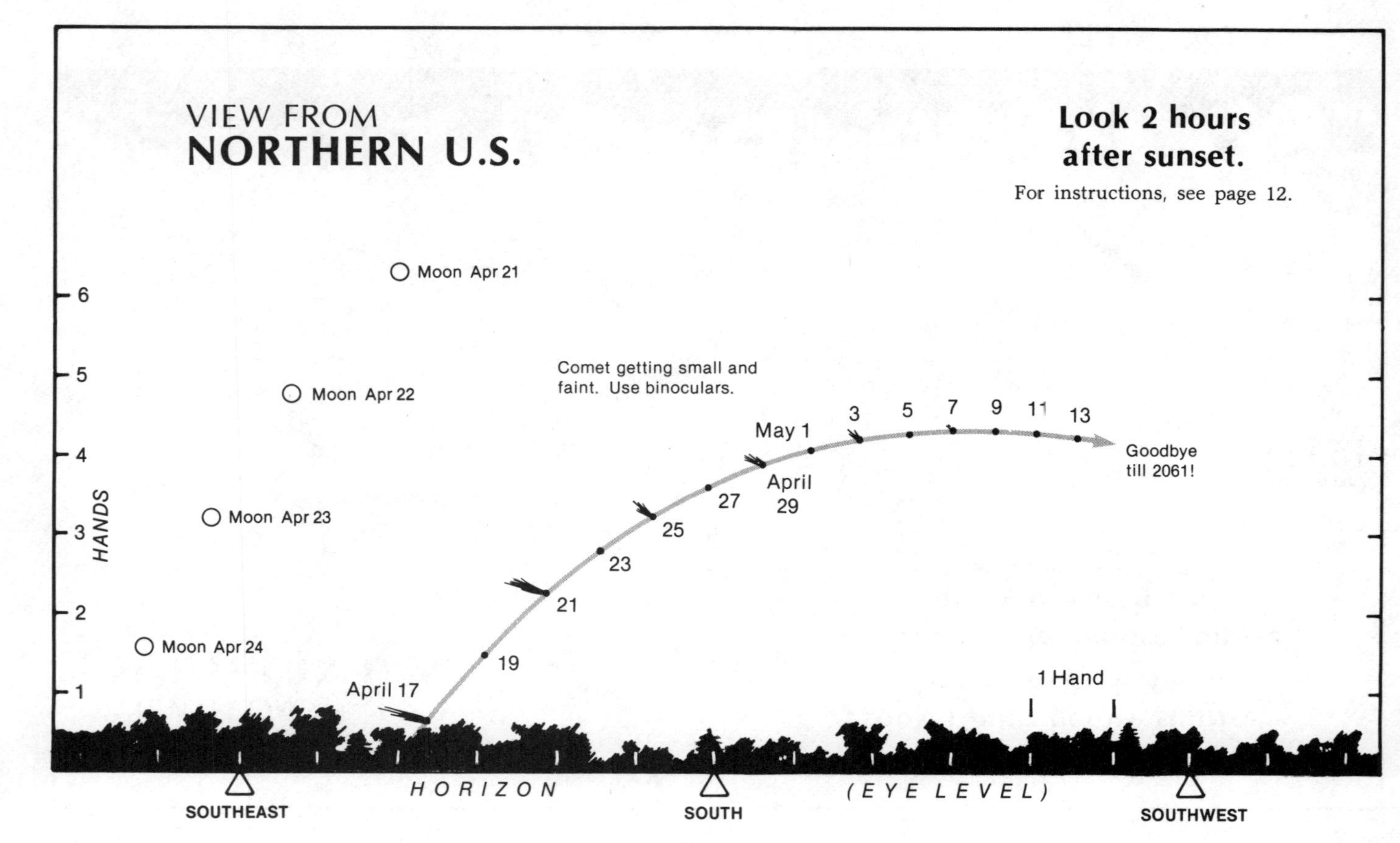
VIEW FROM
NORTHERN U.S.
Look 2 hours
after sunset.
For instructions, see page 12.
Moon Apr 21
Moon Apr 22
Moon Apr 23
Moon Apr 24
Comet getting small and
faint. Use binoculars.
April 17
19
21
23
25
27
April
29
May 1
3
5
7
9
11
13
Goodbye
till 2061!
1 Hand
HANDS
6
5
4
3
2
1
HORIZON
(EYE LEVEL)
SOUTHEAST
SOUTH
SOUTHWEST

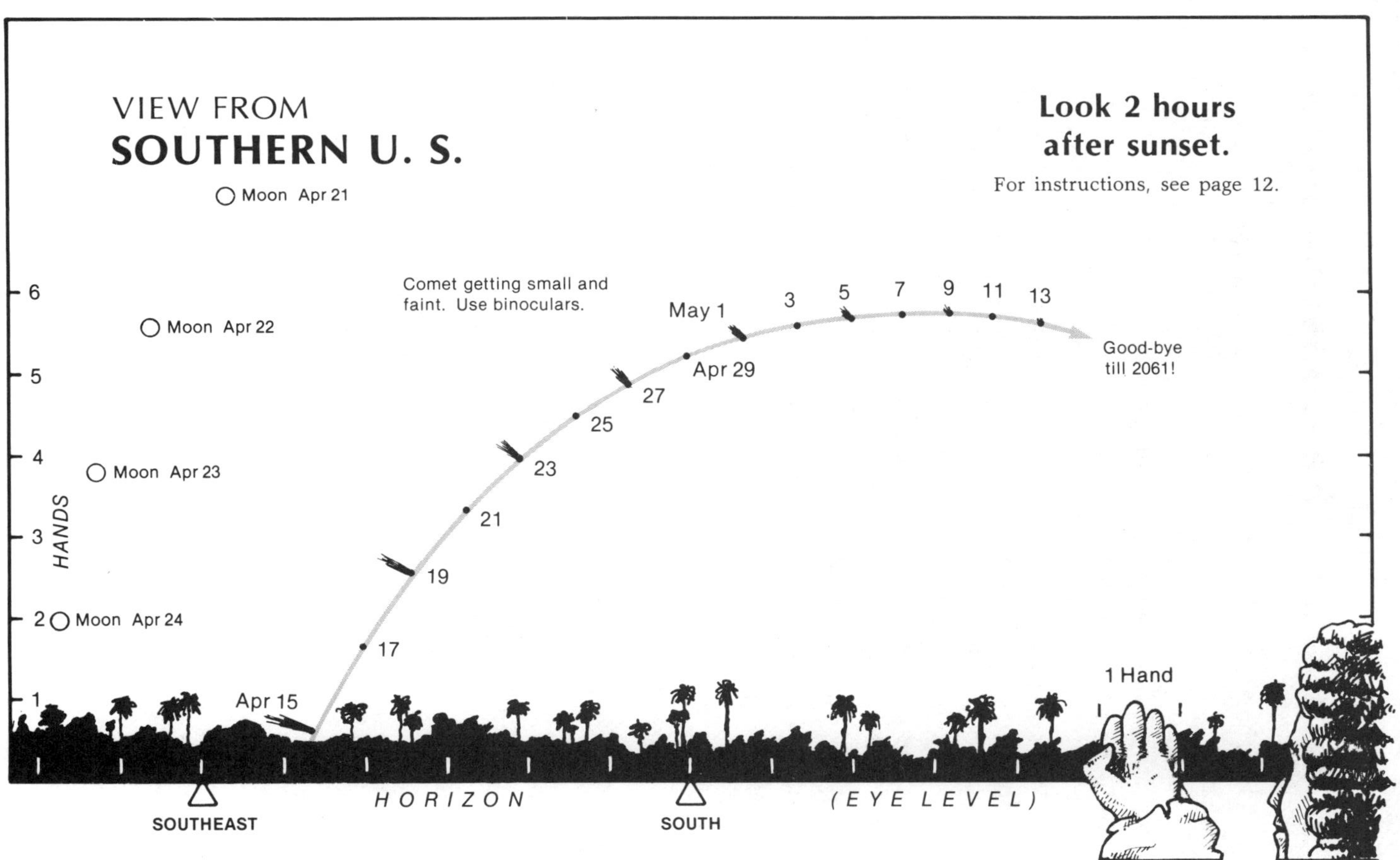

VIEW FROM
SOUTHERN U. S.
Look 2 hours
after sunset.
For instructions, see page 12.
Moon Apr 21
Moon Apr 22
Moon Apr 23
Moon Apr 24
Comet getting small and
faint. Use binoculars.
Apr 15
17
19
21
23
25
27
Apr 29
May 1
3
5
7
9
11
13
Good-bye
till 2061!
HANDS
1
2
3
4
5
6
1 Hand
HORIZON
(EYE LEVEL)
SOUTHEAST
SOUTH

If You Know the Constellations. . .

The charts on the previous pages are all you need for dates when my comet can be seen with the naked eye. But if you have binoculars or a small telescope, you may want to try for it on other dates when it's fainter. To do this, you'll have to know exactly where among the background stars to look.

If you already know the star patterns that form the constellations, or if you know somebody who does, the star map below will help you locate Comet Halley precisely on any date from September, 1985, to August, 1986, at any time of night.

If you don't know the constellations, never mind, you can turn the page and pass this part by. But learning to identify the star patterns in the night sky can be fun! See "If You Want To Learn More" on page 29.

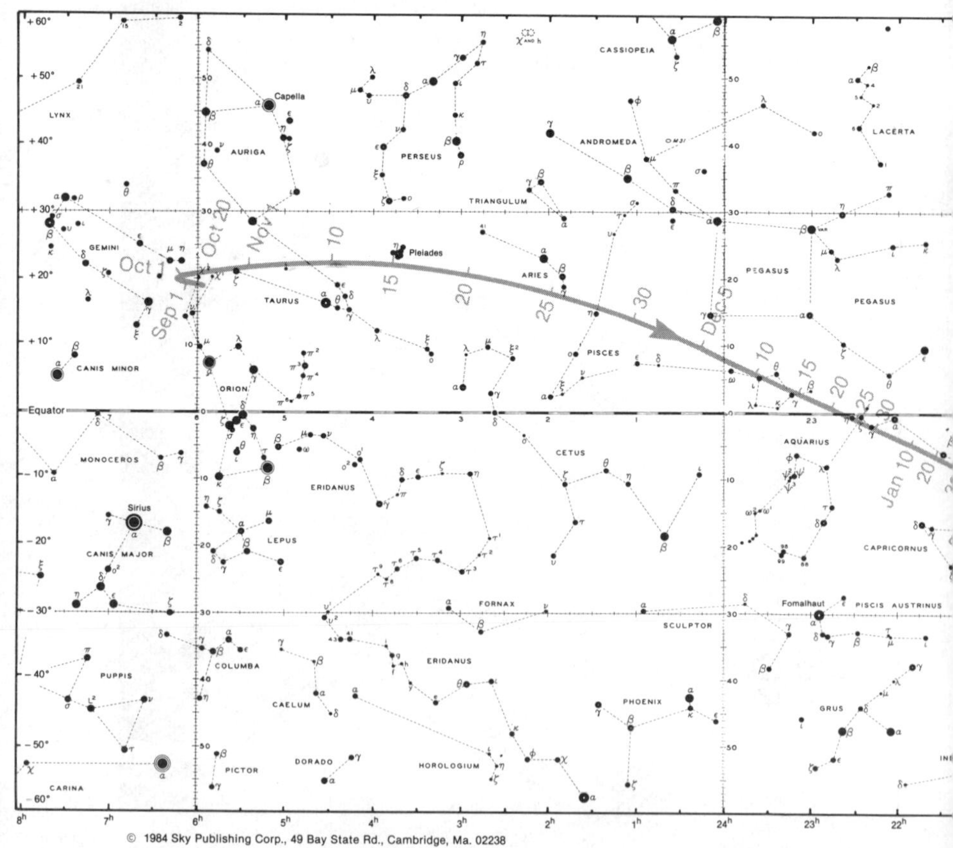

Path of Halley's Comet Through the Stars

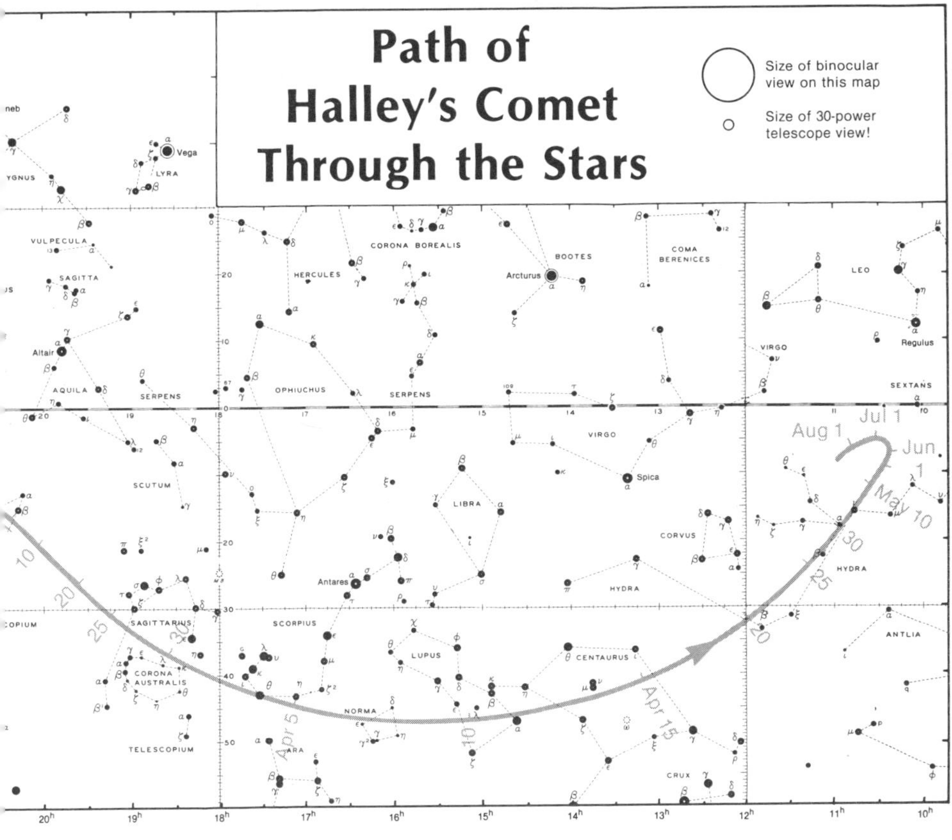

Size of binocular view on this map

Size of 30-power telescope view!

Using Binoculars . . .

The handiest optical aid for looking at the sky is a pair of binoculars. They're a lot easier to use than a telescope and show you a larger part of the sky at once. This is a big advantage for finding Halley's Comet. Any binocular will give a better view of the comet than the naked eye. Use whatever pair you may have around the house.

If you decide to buy binoculars to view Halley, be sure to get a reputable brand; this is your best assurance of quality. For best performance on the comet — and for sky-watching in general — the front lenses should be as large as possible. Magnifications between six and nine power are fine for comet viewing.

. . . and Telescopes

Halley's Comet is prompting lots of folks to buy telescopes. Some people will put their new telescopes to good use: They will follow the comet from night to night, examining its detailed structure and seeing how it changes. These budding amateur astronomers will also begin exploring the sky for other interesting sights.

But other people, it's sad to say, won't see a thing through their new telescopes. They'll fail because they didn't know where to look, or because they didn't learn how to use their new instruments properly.

Such oversights are strange, because these same people know you can't become an expert cook just by buying pots and pans. Gaining any skill takes study and practice. It's the same with becoming an amateur astronomer. A telescope is just a tool; you have to know how to use it correctly.

Here are some tips for looking at my comet.

• Use the telescope's *lowest* power (magnification). Low power gives the widest view, making the comet easier to find and showing more of it at once.

By the way, don't pay attention to ads that boast about a telescope's super-high power. The important thing about an astronomical telescope is its *aperture,* the size of its main lens or mir-

ror. The larger the aperture, the more light the telescope collects and the better the view. A telescope that is advertised for its power and not its aperture is probably not worth buying.

• Don't expect to see a whole lot right away. The image in a small telescope isn't anything like the photographs taken at large observatories. You'll probably see the coma — rather bright, small, and fuzzy — and when the comet is brightest, streaks of the dim tail.

• Be patient. Examine the comet carefully and slowly, giving your eye plenty of time to discover faint details overlooked at first. You'll be surprised at how much more can be seen if you just take the time to study the image thoroughly.

If you're thinking of buying equipment, read the articles "Observing with Binoculars" and "How to Choose a Telescope" in *Sky & Telescope* magazine for October and December, 1983. Check your library, or send $2 to "Binocular/Telescope reprint," *Sky & Telescope,* 49 Bay State Rd., Cambridge, Mass. 02238.

Take a Picture of My Comet!

How I wish we'd had photography in my day! It would have helped astronomers like me find out so much more about the universe, in many ways. And I could have taken pictures of stars and comets to hang on the wall or show to friends. So don't miss this opportunity to take a picture yourself! Here are complete instructions.

To begin with, you must have this equipment:

- A camera that can take time exposures — one that can keep its shutter open for as long as you wish. (On many cameras this is the "B" setting. Some of today's automatic cameras can't do this.) Use a lens of about 50mm focal length.
- A tripod or other support to hold the camera rock steady.
- Fast film. Any photography shop can supply you with one of these films. If you want color prints, use Kodak VR 1000 or Fujicolor HR 1600; if you want slides, use 3M Color Slide 1000.

First, load the camera; and while it's still daylight, take a picture of something. (When your film is developed, this picture will show the technician where to begin cutting the film.)

Now mount the camera on the tripod, focus at infinity (when very distant objects appear sharp in the viewfinder), and set the f-stop to the lowest number (like f/1.4 or f/1.8). Adjust the exposure control to permit time exposures. If you have a cable shutter release, attach it to the camera.

Once you've found the comet with the naked eye, frame it in the camera's viewfinder. You might want to include some foreground scenery such as trees or hills. Press the shutter button (or the cable release) and hold it down for 15 seconds. Be careful not to jiggle the camera! Then advance the film to the next frame, check to make sure the comet is still correctly positioned in the viewfinder, and take another exposure — this time for 30 seconds. Repeat the procedure for an exposure lasting a minute. In a really dark sky, a two-minute or even longer exposure might give the best results of all. If your picture taking has to be done in twilight, take exposures of 3, 7, 15, and 30 seconds instead.

When you bring the film in to be developed, be sure to say you want every frame printed (or made into a slide). At first glance, your comet pictures may look strange to a technician — like there's nothing on them. But at least one of your exposures will almost certainly come out nicely.

On longer exposures, the star images will appear as short streaks. They do so because the Earth turned while you had the shutter open. Astronomers prevent these streaks by using "clock drives" to compensate for the Earth's motion. Mounting your camera on a clock drive is the next step in sky photography, if you want to get more involved.

Halley Space Missions

Halley's Comet has come back 29 times since the first definite record of it in 240 B.C. On most of those occasions the comet was greeted by ignorance and fear. But during its last four returns, fear has given way to scientific study. The last time Halley came back, in 1910, astronomers were able to photograph it for the first time and analyze its light to tell what it is made of.

That year the first automobiles were beginning to replace horses, and a few years before, the Wright brothers had made their first short airplane flight a few feet above the ground. Nobody dreamed that at the next return we would send spacecraft flying across the solar system to *meet* the comet. But that is happening! In March, 1986, five unmanned space probes will carry cameras and other instruments right through Halley's glowing head and tail for closeup looks.

The largest of the five is being sent by 11 European countries. The spacecraft is named Giotto, after an Italian artist who painted Halley's Comet following its return in 1301. Giotto will race through the comet's hazy head at a speed of 42 miles a *second.* In the four hours before closest approach, it will send back pictures of the solid nucleus — the dirty snowball in the comet's core — and much information about the gas and dust the spacecraft is flying through. Giotto will probably be destroyed by hitting dust and pebbles near the nucleus, but not before it radios back what it has found.

The Soviet Union is sending two spacecraft, named Vega 1 and 2. They, too, will take closeup pictures of the nucleus and analyze the comet's material.

Japan's space probes are named Planet A and MS-T5. They will pass farther away from the nucleus than the other probes and will photograph and analyze the gases in Halley's outer regions.

American scientists had also hoped to send a spacecraft, but it was canceled for lack of money. They will, however, study the comet from afar using the telescopes on the Space Shuttle.

In addition, astronomers all over the world will observe the comet with equipment on the ground. This work will be very important in making sense of what the spacecraft find. All in all, this comet will be studied more than any other in history.

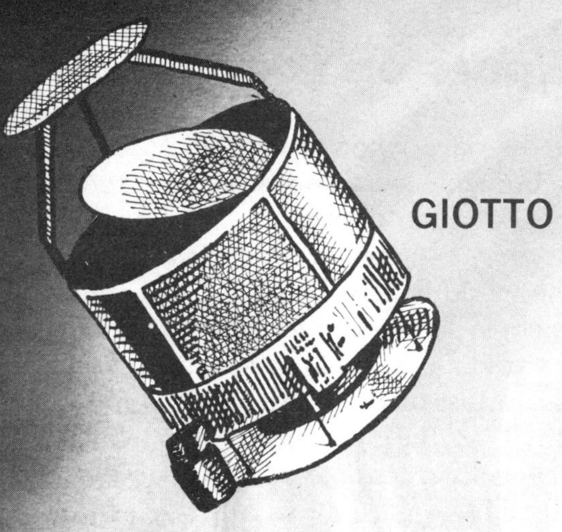
GIOTTO

PLANET A

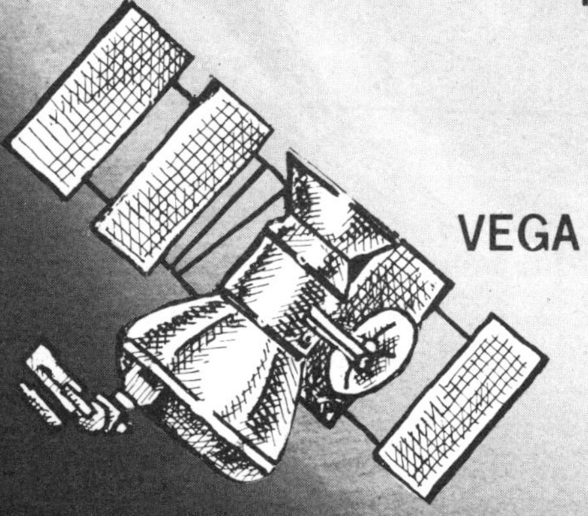
VEGA

Warning: Crackpots on the Loose!

We've learned a lot since the days when people were scared of comets and thought they brought doom and disaster. Unfortunately, a few folks still believe such nonsense.

Even today, groups are trying to scare us by saying the appearance of Halley means the world is going to end. For centuries, sincere but misguided people have claimed the next comet will signify the end of the world. But, in fact, several comets pass by the Earth every year. Most of these are small, dim objects that don't make headlines, so the doomsayers don't know they're around. Only comet celebrities like Halley put these peoples' heads spinning. Already, some groups are buying big newspaper ads to predict a ''fiery comet chastisement'' for mankind. Some folks just never learn.

Millions of miles away in space, Halley's Comet has nothing to do with us. It's going its own lonely way. We're lucky to be living at just the right time to see this once-in-a-lifetime sky event. So relax and enjoy the show.

Kurtze Erklerung/von den
eigenschafften deß grossen/im M.D.LXXVII.
jars erschienen vnd noch brinnenden Cometen/Samt sei-
ner bedeutung/ Auch an was ort vnd stell/ Himlischen Hauß/ Pla-
neten vnd Zeichen/er auff vnd vnter gangen/ Durch der Astro-
nomiæ Liebhabern/ erfahrne vnd Gelehrte/ wie nach-
uolgt/vnderschiedlich beschrieben.

Comet craziness, then and now. Skeletons predicted death and destruction in this German flyer about the comet of 1577. An American tract on Comet Kohoutek announced that the world would end on January 31, 1974.

If You Want To Learn More

Comet books

The Comet Is Coming! by Nigel Calder. 160 pages, paperback, $6.95.
The history, legends, and science of Halley's Comet through the ages. Lots of pictures. Reading level: general public.

The Return of Halley's Comet by Patrick Moore and John Mason. 121 pages, hardbound, $14.95.
Detailed information about the comet, past and present, by two well-known British astronomy authors. For the general public and amateur astronomers.

Current news

The Planetary Report, published bimonthly by the Planetary Society, is a beautifully designed and illustrated publication directed toward readers interested in planetary science, but who may have no scientific training. It covers all aspects of the subject, with emphasis on spacecraft and human exploration of the planets. It is available only as part of membership in the Planetary Society, 110 S. Euclid Ave., Pasadena, Calif. 91101.

Sky & Telescope, published every month, is the world's leading astronomy magazine. It contains articles at all levels from beginner to advanced. Gives much information on current sky happenings and will have full updates on Halley's Comet. Subscription: $18.00 per year in the United States, $21.50 in Canada and Mexico. See address below.

Learning the sky

The Stars by H. A. Rey. 160 pages, paperback, $8.95.
An excellent, simple, very clear beginner's guide to the stars and constellations. If you want to start learning the sky, this is the book to use. Reading level: teenage and up.

Using a telescope

The Edmund Mag. 6 Star Atlas by Terence Dickinson, Victor Costanzo, and Glenn Chaple. 66 oversize pages, paperbound, $11.95.
All the basic information the owner of a new telescope needs to know. It assumes you can already find your way around some of the constellations. High-quality star maps pinpoint hundreds of celestial sights for telescopic viewing: double stars, variable stars, star clusters, nebulae, and galaxies.

All the items above can be ordered through Sky Publishing Corporation, 49 Bay State Road, Cambridge, Mass. 02238. Include full payment. (No charge for postage.) Massachusetts residents: add 5% sales tax. (Foreign orders: add 10% for shipping. Payments must be in U. S. funds drawn on a U. S. bank or by International Money Order.)